MATH ATTACK: HOW TO REDUCE MATH ANXIETY IN THE CLASSROOM, AT WORK AND IN EVERYDAY PERSONAL USE

BY

MARILYN CURTAIN-PHILLIPS, M. Ed.

ISBN 0-9673997-1-8

Printed in the United States by:
Morris Publishing
3212 E. Hwy 30
Kearney, NE 68847
1-800-650-7888

MATH ATTACK REVIEWS

" An interesting and informative treatment of an important problem in contemporary education. The work presents many useful strategies for dealing with "math anxiety" and, in my judgment, would be a particularly valuable source for those planning to teach mathematics at the introductory or high school level."

Michael Becker, Ph.D., Mathematics Department
University of South Carolina

"As a first year teacher, *Math Attack* is a book that I would recommend to another novice in the field. In the future I plan to use a great deal of concepts brought up to help students enjoy math more. Being a novice, there were many things that I was not aware of, but thanks to *Math Attack*, I hope to offer my students some ease so that they can give more back to me."

Pamela D. Rogers, Mathematics Teacher

"... a nice supplement to a mathematics teacher education course. A broad bank of ideas for educators..."

Jane Plutt, Math Tutor
Opportunity Scholars Program
University of South Carolina

"By reading *Math Attack*, it helps the way I do math. It made me feel math wasn't hard at all. At first I had a 72 in my math class, after I read *Math Attack* I finished my class with an 85."

Beanthay Peoples
High School Student

"*Math Attack* gets inside the secret world of causative problems afflicting millions of young and adult people suffering from math phobia, and offers holistic information that can revolutionize math-friendly education in America."

Loretta Pollard, Lay Educator-
School Board Member

"I have been a professor of chemistry for more than twenty years, and have observed, repetitively, that poor mathematics preparation inhibits success in the sciences. Poor student preparation is due mainly to self imposed anxieties, which inhibit the joy of learning mathematics, and the failure of institutions to address and accommodate diverse learning styles. Book topics are excellent reading and discussion for pupils who are in early and beginning level arithmetic courses. This book is also useful for adults who are retooling for more competitive technical careers."

Albert N. Thompson, Jr., Ph. D., Chemistry
Department, Spelman College

"In this dawn of a new century, technology is expanding. There is a greater need for mathematics in everyday situations. I believe *Math Attack* will assist those nervous about the subject of mathematics and help to expand their horizons."

"I found the section on Test Taking Tips and 77 Careers & Math Needed to be very useful. Many of the tips that are in the book have helped me over the years. Knowledge of which careers involve math and the years of math needed helps students realize the importance of a strong mathematics foundation. I would recommend this book to be used in High Schools for the beginning math classes. I also recommend it for those out of school who shied away from mathematics-laden courses. If reached early enough math anxiety can be reduced and conquered."

"In my profession mathematics is used extensively everyday. However, I was recently approached by a coworker, who wanted to build a slide for her daughter. I was able to assist her using the Pythagorean Theorem now that's math being used in everyday life."

Joyce M. Rose-Harris
Actuarial Assistant

ACKNOWLEDGMENTS

I would like to give my deepest thanks to my family and friends for all the support I received throughout this process.

In loving memory of my father, Ernest Albert Curtain, who taught me to always do my best.

In loving memory of my mother, Annie Furgess Curtain, who was a great math teacher and an even greater mother.

Great thanks to God, who makes all things possible.

PREFACE

This book provides strategies to reduce math anxiety in the classroom, at work and at home. Math anxiety is a fear of uneasiness about math that causes many to avoid math or panic when dealing with math.

The main emphasis of this book is to focus on methods to overcome math anxiety by examining learning styles, alternative teaching methods, and the impact and causes of math anxiety on society.

The goals of this book are to improve student feelings toward math, student achievement, student confidence and self-esteem, provide teachers with different teaching methods and give parents tools to help their children become successful in studying mathematics. Feedback from students exceeded the author's expectations in a short period of time. The student's confidence increased. Students, who on the first day of class said they did not like math, now volunteer to present problems to class and participate more actively in class.

TABLE OF CONTENTS

INTRODUCTION

This book was prompted by an educator's observation that many students have mathematics anxiety. These students are constantly frustrated due to poor grades, poor math related experiences and lack of understanding of mathematics. Therefore these students do not enjoy mathematics and do not see the need for its usefulness in their lives. Many students are placed in remedial math classes. In the traditional classroom setting, students are left with the feeling mathematics can only be understood by a certain few. This feeling of inadequacy remains with many students throughout adulthood, causing them to choose careers that limit the use of mathematics.

History of Mathematics

"Mathematics is a necessary foundation for most subjects students study at a modern university: natural science, social science, technology, engineering, agriculture, management, business" (Tobias, 1987). The need for mathematics is what this author stresses to her students throughout the school year. Students are challenged to think of a career which does not use mathematics. Students name a career then the author and or other students will state how math is used for that particular career. Later in the school year students will research a career, its education requirements, duties and develop ten math word problems.

"Mathematics is, by nature, a study in which old ideas are continually replaced by new ones, and the importance of a new work may be measured by the number of former ones that it renders obsolete. Mathematicians are normally concerned with the present and the future. The past is dead and buried"(Dubbey, 1970).

"As the society's needs changed so did the evolution of mathematics. Cave paintings date from 15000 B.C., and it may be deduced from this that primitive tribes have reached a similar level of art, and also possess a simple arithmetic. Elaborate number systems were used from the Late Stone Age, c.5000 B.C. It was necessary to count and calculate in order to trade, to keep large flocks, and to make war" (Dubbey, 1970).

"The Egyptians were the best engineers, technicians, chemists and physicians of the ancient world, and the first to recognize that the year consisted of 365 days. They built the Pyramids in the period from roughly 3000 to 2000 B.C." (Dubbey, 1970). The Egyptians used a system based on groups of 10 and developed basic geometry and survey techniques, evidence concerning the origins of this subject.

"By 2100 B.C. the people of ancient Babylon had developed a sexagesimal-system - a system based on groups of 60. This system is used today to measure time in hours, minutes and seconds" (Dauben, 1994).

Pythagoras (c.572-c.497 B.C.) of Samos, is one of the best known of Greek mathematicians. He is most known for the Pythagoras' theorem. His more useful contributions to mathematics were probably in the classification of numbers and the introduction of geometrical algebra. Pythagoras founded a sect which based itself on strict secrecy and aseptic living for the purpose of advancing knowledge by means of the study of numbers as the key to open the mysteries of the universe. They classified numbers into odd and even, prime and composite.

Euclid, one of the foremost Greek mathematicians, wrote the Elements about 300 B.C. In this book, Euclid constructs an entire system of geometry by means of abstract definitions and logical deductions.

After the decline of Greek civilization, the intellectual emphasis in the Western world was on theology. The studies of mathematics began to slow down. The first outstanding mathematical book produced in the West was the Liber Abaci of Leonardo of Picsa (Fibonacci) in 1202, a work which remained authoritative for over two hundred years according to Dubbey. Practical applications helped to advance mathematics during the 1200 in Italy. The twelfth century saw the construction of many of the great European cathedrals with the need for a practical knowledge of geometry.

The Renaissance also brought major advances in pure mathematics. Between the years 1472 and 1500, over two hundred mathematical books were published in Italy, mainly to satisfy needs arising from shops, banks, mines, armies, universities and astrologers. The leading mathematicians took part in public problem-solving contest, and developed their own techniques, usually in secret, in order to excel at this unexpected forerunner of prize-fighting (Dubbey, 1970).

By 1600, the increased use of mathematics and the growth of the experimental method were contributing to revolutionary advances in knowledge. Sir Isaac Newton and Gottfried Wilhelm Leibniz published their independent discoveries of calculus in the mid-1680's, (Dauben, 1994). Progress in the advancement continued throughout the 1700's to 1800's. Contributions were made by mathematicians Jakob Bernoulli, Leonhard Euler, Joseph L. Lagrange and many others (Dauben, 1994).

In the 1800's, public education expanded rapidly, and mathematics became a standard part of university education. Many of the great works in mathematics of 1800's were written as textbooks (Dauben, 1994).

"New areas of mathematical specialization have arisen during the 1900's, including system analysis and computer science. Advances in mathematical logic have been essential to the development of electronic computers. Computers, in turn, enable mathematicians to complete long and complicated calculations quickly. Since the 1970's, computer-based mathematical models have become widely used to study weather patterns, economic relationships and many other systems" (Dauben, 1994).

Math has been seen throughout the ages as something that only a certain few could do. Math developed due to the necessity of the times. The negative feelings associated with mathematics have been around for quite sometime and remains today.

This author recalls college days when asked about one's major, students began to confess their dislike of mathematics. These confessions came from elementary education, science and English majors to name a few. Math majors were seen as brains. These negative feelings toward mathematics came from people of various backgrounds, race and gender.

This author believes the negative feelings toward mathematics has caused many individuals to have math anxiety. Individuals who have done well in many other fields of studies but perform poorly in mathematics because of math anxiety.

"Forces created by computers, applications, demographics, and schools themselves are changing profoundly the way mathematics is practiced, the way it is taught and the way it is learned" (Steen, 1990).

School mathematics has changed greatly since this author attended elementary and high school from 1966 - 1977. This author recalls math classes which were simply pencil and paper computations with textbooks. A slide ruler was the greatest instrument available in high school. Calculators were now available but extremely expensive. These calculators were shun by most math teachers and not allowed in class. The calculators were believed to weaken math students' skills. Memorization was the most essential skill to have in a math class.

Clearly school mathematics has changed greatly since those days. A slide ruler is no longer used. Calculator use is encouraged by many school administrators and teachers are provided with a classroom set of calculators. The cost of calculators have decreased greatly. Calculators are priced as cheaply as $2.00. A scientific calculator can be purchased for $5.00.

Computers are now available to students. There are computers in school libraries, computer labs and in classrooms. The Internet has made information available around the world from large cities to small towns within seconds.

As early as 1957, Dreger and Aiken suspected that individuals suffered from number anxiety and discovered that this newly created construct correlated with final mathematics grades (Measurement & Evaluation in Counseling & Development, April 1996). Mathematics anxiety has been defined as "feelings of tension and anxiety that interfere with the manipulation of numbers and the solving of mathematical problems in a wide variety of ordinary life and academic situations

(Richardson and Suinn, 1972). Researchers have found that mathematics achievement (e.g., Pajares & Miller, 1994, 1995a: Ramirez & Dockweiler, 1987, mathematics aptitude (e.g., Cooper & Robinson, 1991; Hackett & Betz, 1989), affects grades in mathematics classes (e.g., Wigfield & Meece, 1988).

From personal experience this author can recall talking to friends who experience math anxiety. One friend, a highly skilled police detective who has received national attention, recalls having math anxiety. The police detective recalls, when she got ready to take a job placement math skill test, she panicked. She said she couldn't remember what to do. She knew she had seen these types of math problems over the years in middle school and high school. However, due to the pressure of a testing environment and the grade a deciding factor in her career, she experienced a mental block.

However, this same investigator can go into a homicide scene and tell what angle a bullet must have entered the body, due to the amount of impact. When math is used on her job in a practical problem solving application, calculations can be done with ease. Her focus is then with the problem at hand and away from math itself and the fear of failure. She sees math as a tool, an aid to accomplish a task, no longer an obstacle as she did in the earlier incident.

Sepp and Schwarzer analyzed the relationship between mathematics anxiety and math related performance outcomes from 1975 to 1986. Over the years this author has had students who had math anxiety. One student did extremely well orally and on class work assignments. She volunteered to answer questions. But during test time her grades were very low. This author pulled the student aside and asked her what was wrong. The student said seeing the words quiz or test at the top of the paper caused her to get nervous and to forget what she did earlier in class. The author suggested the student fold back the part of the paper that said quiz or test and then begin. her grades changed greatly from D's to B's. Something this simple had a great impact on this student's grade and comfort level during testing.

Major Goal

Research confirms pressure of timed tests and risk of public embarrassment have long been recognized as sources of unproductive tension among many students. Three things that cause great anxiety in many students are imposed authority, public exposure and time deadlines. The things which are a regular part of the traditional mathematics classroom cause great deal of anxiety. Therefore, teaching methods must be re-examined. There should be more emphasis on teaching methods which include less lecture, more student directed classes and more discussion.

Given the fact that many students experience math anxiety in the traditional classroom setting, the major goals of this book are to ensure a high success rate and arrange problems so that students will seldom fail. Teachers should ask more open ended questions. These types of questions will encourage different types of responses that can be discussed rather than just one right answer. If a student gives an incorrect answer, teachers should use this experience and respond by saying, "That's a good point but that would be used with..." An incorrect response can be used as a quick review time of an earlier concept. The class would return to the question at hand. Students must have a high level of success or a level of failure that they can tolerate. Incorrect responses must be handled in a positive way to encourage student participation and enhance student confidence.

While researching for this book, the author will determine through a survey whether these students are engaged in mathematics class and analyze their learning styles. The survey is a questionnaire regarding student's personality. Students will check whichever statement applies to them individually. Statements will include the following: Do you have a lot of friends? Do you socializes a lot at school or at home? A series of lessons will be presented showing the traditional method of instruction. Then present math lessons using increased use of manipulatives and open-ended questions and alternative assessment strategies. Students would engage in cooperative activities, experimental, challenging games and creative art work. This will then lessen math anxiety of students in math classes, giving them opportunity to let their various talents shine through in ways often unnoticed in the traditional math class. Finally, these students will be surveyed again to determine whether the different means of lesson presentations have had a direct effect on their feelings toward mathematics.

Rationale

The purpose of this book, therefore, is to provide students, parents and the public with an opportunity to gain a better understanding of mathematics. There are numerous books and magazine articles written regarding math anxiety. One definition given for math anxiety is the "emotions" clutter one's understanding and recall of ideas as one attempts to solve math problems. Math anxiety can cause one to forget and lose one's self-confidence. Studies have shown that students learn best when they are active rather than passive learners. The theory of multiple intelligences addresses the different learning styles. Lessons are presented for visual/spatial, logical/mathematics, musical, body/kinesthetic, interpersonal and intrapersonal and verbal/linguistic. Everyone is capable of learning, but learn in different ways.

This society no longer has a great need for factory workers, but instead has a greater need for technical workers. Today's workers are needed in the areas of business, engineering, science and computer science. All these areas require workers with a strong mathematics background. Employers have expressed the frustration of not being able to find qualified applicants. The need for mathematics is becoming greater than ever before. Thus, presence of math anxiety is having a greater impact on society. It is essential to address the issue of math anxiety such as its causes, symptoms and effect and provide strategies on how to overcome it.

Population

The population to be served during this research process were the author's 9th and 10th grade math students. These students were from a small rural town in South Carolina and had been placed in "Mathematics for Technology", a remedial math course. This course was designed for students who generally do not plan to attend college but for those who enter the workforce upon graduating from high school. The classes consist of mostly African American male students.

LITERATURE REVIEW

A great deal of literature has been written with regards to the existence and impact of math anxiety. "There have been many changes in teaching in the last forty years. These changes encompass the content of what is taught, the method of teaching it, and the relationship between teaching and the degree of control asserted in the classroom"(Buxton, 1991).

There are many factors which have changed teaching methods in the classroom. More studies have been done in regards to learning styles. Teachers are now aware all children do not learn the same way all the time. Lessons must be presented in a variety of ways. The different ways to teach a new concept can be through play acting, cooperative groups, visual aids, hands on activities and technology. Learners are different than they were forty years ago. These learners ask questions why something is done this way or that way and why not this way. Whereas, years ago learners did not question the why of math concepts; they simply memorized and mechanically performed the operations needed.

Students today have a need for practical math. Math needs to be relevant to their everyday lives. Students enjoy experimenting. Math fairs provide students with the experiences of making presentations and performing hands on activities.

Projects and fairs are an excellent way to get parents involved.

Adrien Hess stated fairs and competitions in all areas of industry and art are familiar features in American life. Such experiences are pleasant and educational for all concern-the exhibitor learns a great deal in preparing the exhibit and the viewer learns something new from seeing it (Hess, 1992). There are now books and articles available on how to have a math fair.

"Children learn best when they are active rather than passive learners" (Spikell, 1993). "Knowing mathematics means being able to use it in purposeful ways. To learn mathematics, students must be engaged in exploring, conjecturing, and thinking rather than only in rote learning of rules and procedures. Mathematics learning is not a spectator sport. When students construct personal knowledge derived from meaningful experiences, they are much more likely to retain and use what they have learned. This fact underlies teachers' new role in providing experiences that help students make sense of mathematics, to view and use it as a tool for reasoning and problem solving" (National Council of Teachers of Mathematics, 1989).

Each day this educator's students come in and complete warm up math problems or brain teasers at the very beginning of class to get students focused. From each group, a member presents a solution to a problem and earns team points. If another team can solve the same problem in a different way, they can also present and earn team points. Students enjoy this activity and sees it as a challenge to find different ways to solve a math problem. For the author it is enjoyable to see the different ways that children think. This author believes the more children are encouraged to approach math problems in different ways the more they will have confidence in their own math abilities.

Many students are trapped into the belief there's only one right way to solve a math problem. During the time of a test or in front on a class, a student might forget the one "right" way he believes to solve a problem, anxiety will then set in. Students enjoy a student centered class. Many students enjoy taking on a leadership role, presenting their ideas, explaining their method used and explaining why. Classmates are less hesitant to ask a classmate a question and they feel free to make comments or suggestions to the presenting student.

This student interaction and student led discussions are a part of the reform called for in math education. The teacher becomes a mediator and assists students with their findings. The class goes in the direction of interest given by the students rather than a rigid lecture.

Students are used to a faster pace and more communication with others. The author believes it is important for students to learn a sense of team work and cooperation. These are essential skills needed in today's workforce. Workers must be able to get along with others. Most workers' duties are dependent and related to someone else's duties.

All teachers of mathematics from kindergarten through high school face the challenge of meeting the national call for reform in math teaching. Students who understand mathematics can think and reason mathematically and use what they've learned to solve problems, both in and out of school. Teachers who teach for understanding must find ways to engage students actively in their mathematics learning (Burns, 1992).

Students are naturally curious. They enjoy puzzles, games and mysteries. Math word search puzzles are a wonderful way to review vocabulary words. Word search puzzles help improve students' spelling of math terms and also is a time for discussion. Students will ask questions about a word they are not familiar with but are trying to find within the puzzle. Math With Pizzazz series have great riddles that can be uncovered by solving various types of math problems.

Classroom visuals can be used to stimulate students' curiosity as a preview of topics to come and as a review of previous math concepts. Bulletin boards can be made interactive- students solve problems on the bulletin board.

"In the classroom, math should not be kept hidden inside textbooks and workbooks. Here of all places, it needs to be alive! As a teacher, you can best connect math to real life by making it an active visible part of every student's day. Exciting math bulletin boards are one wonderful way to get math off the worksheets and upon its feet"(Frank,1986). Frank stated in Bulletin Board Ideas for the Math Classroom that all of them are complete math lessons in themselves. Frank also provides suggestions for extending many of the activities.

Peripherals or visual aids address the learning style of visual-spatial learners. These learners would be stimulated by bulletin board themes and posters around the classroom. "These people excel by learning with images, pictures, charts, graphs, diagrams and art. They enjoy visual media that include films, slides, videos, maps, computers, stencils, charts, models, etc. and often spend long periods of time on visually-oriented projects. Such learners frequently arrive at unique, unconventional solutions to artistic problems rather than relying on traditional, conventional approaches (Cassone and Cassone, 1990).

Research by Howard Gardner has indicated seven distinct "intelligences" which are the essential ways by which people learn. Generally only two of the intelligences are taught to"(Cassone and Cassone,1991). They are the linguistic intelligence and mathematical/logical intelligence. For linguistically oriented learners: These individuals have an innate love of language. They often have well developed vocabularies and their fluent use of language includes a richness of expression and elaboration. Reading, writing, editing, listening and speaking opportunities are enjoyed. Working with books and diverse printed materials, records, tapes, lectures, word processors,

etc. are pleasurable activities for these learners."
"For logically-mathematically oriented individuals: Such learners enjoy forming concepts, looking for patterns and relationships, and doing activities in a sequential manner. They like to arrange the steps of a project into a sensible order or timeline and desire time to complete each component. These learners enjoy opportunities to problem solve, learn quantities of facts and have time to explore new concepts. They frequently ask many questions and desire logical and clear explanations. Problem solving, working with games, kits and puzzles, collecting, classifying, organizing information tasks and people are enjoyable pastimes"(Gardner,1990).
"A uniform way of teaching and testing is patently unsatisfactory when everyone is so different" (Gardner, 1990).

The kinesthetic learners are individuals who learn best by moving, touching and doing. They are usually not attentive to visual or auditory instruction, but are eager to attack problems physically and with great activity. Manipulatives, role play, stimulation's, physical exercise, games, competitive sports and action packed stories are enjoyed. These learners will remember best what they have physically done (Cassone and Cassone,1990).

The musically-oriented learners according to Cassone and Cassone, are individuals who enjoy rhythm and melody. They eagerly participate in musical activities. These learners may enjoy soft music being played in the classroom or set a tune to a concept being taught.

The interpersonally-oriented learners are individuals who enjoy learning by interacting and cooperating with others. They eagerly participate in group work and effort, discussions, finding out how others feel, and getting involved in clubs and community service programs (Cassone and Cassone,1990).

It is beneficial for educators to be aware of the different types of learners. It is necessary that lessons be presented in a variety of ways to reach the different types of learners. If a student is always taught in a style that does not ever comply with his or her learning style, this will cause the student great anxiety toward a subject. Therefore, it is important that math classes be taught in different ways. Some days this educators' students come in and may engage in a game of math bingo as a way to review. The teacher holds up a flash card and reads out a question. The students look for the answer on their game card. Students are attentive because they want to win.

Another day this educator's students may come in and engage in team flash card competition. In the competition, questions are asked. The students work in groups. The students discuss within their teams and write down an answer the team has agreed upon. Team answers are held up at the same time.

Some days students will be given manipulatives as a follow up to a concept introduced earlier or as an introduction to a concept. For example, one days students were given different size circular shaped objects. The students measured the diameter across the object then used a string around the object. The measurement around the object was the circumference. The last part of the experiment was to divide the circumference by the diameter. Students were asked what was the pattern. The pattern was that the circumference of a circle is a little more than three times the diameter of the circle. Students then were able to see why (3.14) is used in the formula Circumference = pi x diameter.

The follow up to this activity was to find the circumference of a circle when given the diameter. Students were asked to recall their earlier experience from the hands on activity.

Another day this educator's students may come in and take part in an art activity. While graphing data, students plotted coordinates and made jack-o-laterns during the month of October. The jack-o-laterns were placed around the room. The drawings were great decorations for the holiday season and this gave students an opportunity to display their creative artistic side. Students who were very creative were given a chance to shine. This is important for the student who is not good with computations or shy in class discussions but is really talented in art. Art gives them an opportunity to stand out and feel special.

As a teacher it is important to present lesson in a variety of ways. When students are stimulated there is less time for classroom disruption. Students will ask later about an activity they engaged in earlier when the activity will be done again in class. They will feel more involved in their learning experiences.

The program, "Friendly and Teachable Hands on Math (FATHOM)" focuses on equipping teachers to address a broad spectrum of essential non-arithmetic topics-- with special emphasis on algebra and geometry. By the end of the program's first year, 100 percent of project teachers reported more confidence in addressing the non-arithmetic ares of the math curriculum, increased use of manipulatives and open-ended questions, and alternative assessment strategies. Some admitted they were still anxious, but had made great gains in overcoming their math anxiety. As the mentor teachers became involved in high-quality, non-threatening, hands-on instruction, they began to relax and feel more confident teaching strands of math that some teachers (admittedly) had neglected in the past (Thrust for Educational Leadership, 1995).

"Students' learning is supported when they have opportunities to describe their own ideas, hear others explain their thoughts, speculate, question, and explore various approaches. To provide for this, learning together in small groups gives students more opportunities to interact with concepts than do class discussions. Not only do students have the chance to speak more often, but they may be more comfortable taking the risks of trying out their thinking during problem-solving situations in the setting of a small group" (Davidson,1990).

"Since all mathematics is essentially arithmetic in various guises, we empahsized the importance of a thorough understanding of arithmetic as the basis for understanding all other branches of mathematics" (Motz and Weaver,1991).

Theoni Pappas (1994) emphasized that mathematics is not a rigid, fixed curriculum as believed by many but is full of ideas and encourages readers to discover mathematics where least expected.

Students must be reminded that math is all around. Math exists outside of the classroom. Math is in nature, music, architecture, sports, recreational games, in stores and at home. Math usage is much more common than many people realize. Math affects everyone in some way or another. No one can escape math. Regardless of one's comfort level with math it can not be avoided. This author believes as a math educator it is very important to make math experiences as pleasant and interesting as possible. Math teachers must be aware that math anxiety does exist and must do whatever is possible to help students to overcome this anxiety or minimize it.

"Innumeracy, and inability to deal comfortably with the fundamental notions of number and chance, plagues far too many otherwise knowledgeable citizens" (Paulos,1988).

As mentioned earlier since this educator's college days the dislike of math and math anxiety has been told to this educator numerous times. Math anxiety has been expressed by foreign language, English, history and science teachers to the author over the years. People have excelled in their fields but felt incompetent with mathematics.

One science teacher told the author of her experience of being yelled at by her father while learning her eight multiplication table. She thinks of this whenever she recalls her feeling about math. She sees math as a field she is not good in and does not enjoy. Her feelings of anxiety and tension from elementary school remain even after learning her eight time tables, completing high school and college math courses. Regardless of her later success with math she still associates math with her early years.

Prior negative experiences in math class and similar experiences at home when learning math are often transferred and cause a lack of understanding mathematics (Buxton,1991). "Millions of adults are blocked from professional and personal opportunities because they fear or perform poorly in mathematics. It is not a failure of intellect but a failure of nerve"(Tobias,1993).

As this author earlier mentioned, a student's grades that improved greatly may not be due to change in intellect but a slight change in testing environment. The student said she was studying the same as she did before but was now more comfortable during testing. Her nervousness and insecurity caused by her fear of quizzes and test caused a earlier mental block.

"In this culture, numbers and math are frequently associated with pain and frustration. Unbalanced checkbooks, unpaid bills, unforeseen debts, and unfathomable IRS forms plague many a family. Dealing with money is the common use of numbers and computations in our society, and their relationship may be too often unbalanced towards actions that say numbers are difficult to understand. In front of children, thou shalt not use numbers in vain"(Wlodkowski and Jaynes,1990). Wlodkowski and Jaynes emphasize parents should show their children how numbers are used by them successfully in positive pleasant ways- ways such as cooking, fertilizing, sewing, sports, problem solving in hobbies and home repairs. Parents' statement, "I'm just no good at math," an unfortunately common refrain, promotes self-defeating attitudes.

Math must be looked upon in a positive light to reduce anxiety. A person's state of mind has a great influence on his/her success. Many games are based on math concepts. Games are enjoyed by many people who consider themselves not good in math. These games have met several of The National Council of Mathematics Teachers (NCTM) standards. The games included are Life, Yahtzee, Battleship, Tangrams and playing cards. These games meet the NCTM standards which are (1) Mathematics as Problem Solving (2) Mathematics as Communication (3) Mathematics as

Reasoning (4) Mathematics as Connections (5) Number and Number Relationships (6) Number Systems and Number Theory (7) Computation and Estimation (8) Patterns and Functions (9) Algebra (10) Statistics (11) Probability (12) Geometry (13) Measurement.

Teachers can use these games in the classroom and encourage usage at home. Teachers can make copies of similar board games for students to take home and play with their families.

"Math is potentially great fun, and math skill yields a sense of mastery and self-esteem. The destroyer of the joy in mathematics is not practice but anxiety - anxiety that one is mathematically stupid, that one does not have that special numerical talent. But math talent is no more rare than language talent. The number of great mathematicians and the number of great poets per million of population are roughly the same. Yet people experience math anxiety to a much grater degree than language anxiety. Why? Because their early training has denied them systematic familiarity with the vocabulary, grammar, and spelling of mathematics. Those of us adults who experience math anxiety must resolve not to let this same educational world be inflicted upon our children "(Hirsh,1992). Many schools and textbook publications have prompted Family Math activities to provide parents with ways to help their children learn mathematics at home which involves the family. National Council of Teachers of Mathematics has published the brochure, "Help

Your Child Learn Math," to provide parents with activity ideas, do's and don'ts.

MacNeal emphasizes looking at math as a language and learning how to understand what numbers really mean. "Mathematics is a form of language involving the communication of concepts through symbols" (Sharma,1997).

Many students have difficulty understanding math terms. This author stresses students should put the definition in their own words. Whenever a term is used as part of a definition, they should look up the term within the definition. For example polygon is described as a plane closed figure. Students would be asked the meaning of plane.

Math terms do have their own unique meaning. However, many math terms should be related to every day terms students know whenever possible. The math term quadrilateral which means four sided plane closed figure is compared to other terms which mean four and starts with qua. Other words discussed would be words such as quarter and quartet.

Teacher must keep in mind that language plays a great deal in students understanding of math concepts.

"Mathematics is the key to opportunity. No longer just the language of science, mathematics now contributes in direct fundamental ways to business, finance, health and defense. The need for mathematics in society is greater than ever before" (National Research Council,1989). "Today's factory workers need to use advanced math, have good communication skills, and understand the use of computers"(Narisetti,1995).

"A far greater problem of math avoidance is the exclusion of many working-class people and people of color from the growing number of careers requiring some mathematical background, or at the very least, an entry test on topics in mathematics. There is a growing awareness of the need for studies exploring the interconnections of race and ethnicity with economic and social status as they effect mathematics achievement and restrict access to college education and careers" (Zaslavsky,1994).

This author sees many capable students go the non-college technical route simply because they do not want to take Algebra II and Geometry. Most colleges require these higher level math courses. Family expectation has a great deal to do with the type of courses students will take. If parents are college graduates or want their children to go to college, these children are more likely to enroll in higher level college preparatory math courses.

With all the tension and anxiety, math humor is greatly needed. Mathematicians are often seen as cold, dull people. This author believes students should be shown that this couldn't be further from the truth. Laughter cures many ills in life. Laughter helps relieve stress. This author has presented cartoons to introduce a concept or for class discussion. Students enjoy cartoons and jokes. The newspaper is a great source of occasional math jokes, of which this author is on the constant look out for.

Math humor has been presented as a way to relieve math anxiety. *The Math Curse*, a story book for children, tells the story of a girl who experienced the math curse for three days and then how the curse was broken (Scieszka and Smith,1995). Professor Martin Weissman has bought humor to mathematics in his book, *Laugh With Math*(1992). Weissman's book contains cartoons of classroom settings followed by math problems pertaining to concepts taught in the cartoon story. The unique cartoon approach is aimed at the huge audience of persons of all ages. Detailed step-by-step solutions are given. He describes his program as a mathephobic cure-all.

"The key issue for mathematics education is not whether to teach fundamentals, but which fundamentals to teach and how to teach them. Changes in society, technology, in schools - among others- will have great impact on what will be possible in school mathematics in the next century" (National Research Council,1990).

As mentioned earlier the teaching of mathematics has changed greatly. To be effective, teachers must change with the times. Society requires certain math skills and other skills may be less needed or eliminated altogether. The use of a slide ruler due to technology is a skill no longer needed. The use of calculators calls for a greater numbers sense skills. Estimation skills are needed to do quick calculations in a fast pace society. Multiple choice tests are becoming more and more obsolete. Problem solving and reasoning skills are more desired skills needed today as opposed to rote drill, memorization and guessing skills.

Mathematics is a non-ending science. It has developed according to changes in society. Mathematics is continuing to change today. Many math skills are no longer needed or need to be improved upon. The United States continues to be behind many other countries in student achievements of math and science. American educators must be aware of the need and work toward improving math achievement and restore students confidence in their math abilities. Technology requires good math skills.

LANGUAGE OF MATHEMATICS

Students were given written direction test on the first day of class. This author distributed a written direction test. Students were told to leave the papers face down and not to turn them over until told to do so. There would be 15 minutes to complete the test. Tension could be felt throughout the air. Students stood there with stern faces. Students were then told to turn the papers over. This author walked around the room and administered the test.

Some students who had followed the written direction test correctly, quietly sat there as many of their classmates continued to work. This author smiled at them and nodded. As this author continued to walk around the room, many students began to feel frustrated and confused about the different steps and asked the meaning of the various math terms such as digit, consecutive and descending.

As the time ended, the students who had followed the written direction instructions, to read everything on the page before beginning and to only write their name on the paper sat there quietly.

This author talked to the students concerning their feelings as they took the test. Many said they were confused and frustrated. Others said they felt tense and nervous. The class then discussed the purpose of taking the written direction test. Students said they could see the importance of following directions and how not following directions can cause a great deal of confusion. This confusion then leads many to math anxiety.

This author reminded students that they must read in English, science, social studies and other classes, and that it is equally important that they read in math class as well. This experience is mentioned to students throughout the year to remind them of the importance of following directions. This author helps them to recall their feelings when they did not follow directions.

"When a child is learning a mathematics concept several vital components are involved simultaneously. These are children's: cognitive development; preparation in prerequisite skills for the mathematics concept; preparation in previous mathematics concepts; mathematics learning personality and facility in Language (both native and mathematical); emotional security" (Sharma,1997). Language is very important in aiding a student's understanding of mathematics. Throughout mathematics there are many terms. As stated earlier, students' confusion was caused simply by lack of understanding math terms.

When presenting a lesson to students, this author regularly asks students what a math term means that has just been used in a sentence by the author. This is extremely important when solving word problems and reading directions. One of the greatest problem areas and concern given by students during initial survey was the need for additional help with solving word problems.

Whenever a term is presented, students are encouraged to give the definition in their own words to the class. For example, when discussing the characteristics of triangles the class examines the prefix "tri". Students are then asked to think of different every day words which begin with "tri." Students usually respond with words such as tricycle and triplets. The class then discusses what these words have in common -three of something. The class then looks at the characteristics and meaning of the word triangle which is a three sided figure with three angles. Students enjoy brainstorming activity, thinking of different words. Brainstorming helps students feel a part of discovering the meaning of terms and they are later more able to recall the meaning of a math term when it is related to their prior knowledge of other words. "One-third of 11th graders agree that they don't even know what their math teacher is talking about" (The Mathematics Report Card -- Are We Measuring Up?, 1986).

"Mathematics is a form of language involving the communication of concepts through symbols. In most cases, the mathematics language is the second language an individual encounters after his native language. Among the concepts embodied in this unique language are at the primary level: classification and ordering of objects, quantity representing actions and collections, size, order, relationships, space, form, distance and time.

On a higher level the language of mathematics is helpful in conveying the result was logical thinking and reasoning. It is used in the collection, classification, and organization of data and then in the analysis and interpretation of that information"(Sharma,1997,p.7)

Language of mathematics plays an important role in the understanding of mathematics. Students also enjoy being given a beginning letter of a math term. This educator asks about a word then gives the first letter if students don't know the word. Students then begin brainstorming. This activity sparks children's curiosity and interest. Students enjoy this and the entire class becomes involved and this also provides a review of other math terms. "Mathematics thinking to a great extent is dependent on mathematics language"(Sharma, 1997).

READING OF MATHEMATICS

The reading of mathematics must also be examined to get a better understanding of the structure of mathematics. “Mathematics textbook authors and teachers tend to assume that students will read their textbooks with pencil in hand - that they will go back and forth between the text, the illustration, the examples and even look ahead to the homework problems, as they read. Properly done, reading mathematics is an activity that involves the reader”(Tobias,1987).

Students should read example problems, cover it up, or close the book, leave it, then return and solve later. Students should then open their books and check for understanding.

Reading down is one way to approach a math text. As a student reads about a math concept he should write his own examples regarding the concept to check for understanding. Math sentences must be read through carefully and they do require more extra work than other reading material. It is also recommended by Tobias to read math sentences out loud and with a friend to aid in clarifying and checking for understanding (Tobias,1987).

Tobias also recommends the reading down technique. Reading down goes from an overview of the subject to the examples and the problems, following the sequence in the textbook. Students begin looking at the mathematics itself; the examples, the sample problems, and the solutions to the problem sets. Throughout reading questions should be asked why the author did each of the steps illustrated (p.20).

When reading a math text students should write notes. While reading, students should jot down questions, comments and thoughts while reading the material. Making detailed notes is important in mastering understanding of new material (Tobias,1987). At the beginning of the school year this educator's students are given a textbook inventory sheet to get the students familiar with the parts of their book. This will be beneficial to them to know where the glossary, table of contents and appendix are located to assist them when studying.

Students must be taught how to take notes. Many students only want to write the answer to a math problem. Then when papers are returned to them the numbers have no meaning. This author insists that students write questions. Warm-up notes are turned in weekly. Students are graded for neatness, use of pencil, writing questions, correct answers and detailed work. A math notebook by each student is graded at the end of each nine weeks. Students are reminded that classwork, homework and notes should be put in their notebook and to date each paper. Parents are told about the notebook expectation at the beginning of the year. Parents are encouraged to check their child's notebook periodically to keep them abreast of class activities and to aid them when helping their children study.

WRITING TO LEARN MATHEMATICS

"The use of the writing-to-learn strategy in mathematics classes is one way teachers can implement Mathematics as Communication, the second standard in NCTM's (1989) Curriculum and Evaluation Standards for School Mathematics (Miller,1991,p.516). "Writing in math class has two major benefits. It supports students' learning because in order to get their ideas on paper, children must organize, clarify and reflect on their thinking." "Writing also benefits teachers because students' papers are a valuable assessment resource. Their writing is a window into what they understand, how they approach ideas, what misconceptions they harbor, and how they feel about what they're discovering"(Burns,1995).

Teachers should make sure students understand the two basic reasons for writing in math class - to enhance and support their learning and to help assess their progress. Tell students to include as much detail as possible regarding their understanding of a math concept (Burns,1995).

Students papers can be used for class discussions and activities. Using them in this way reinforces to the students that the teacher values their writing. Hearing others' ideas shows children different ways to approach problems. Students are encouraged to read their papers aloud. By listening to what others wrote, students learned about different methods they could have used. This inspires a few to revise their work (Burns,1995).

"Have students discuss their ideas before writing them. After a discussion, remind children what they heard can be written, as long as it makes sense to them and they can explain it" (Burns,1995). "To help students get started writing, put a prompt on the board such as, "I think the answer is ____. I think this because______."
What's important is that their writing, no matter how they express it, relates to the problem and makes sense. Students are encouraged to describe their thinking with words, numbers, and if they like pictures" (Burns,1995).

Reading class sets of assignments gives teachers an overview of how the class responded to particular lessons and helps teacher to evaluate the effectiveness of instruction provided. It also gives information on each child's understanding (Burns,1995).

Writing and discussions mentioned earlier are specifically beneficial to the verbal/linguistic intelligence, interpersonal and intrapersonal intelligence. It appeals to the verbal/linguistic intelligence for learners who gain a better understanding by hearing, speaking, reading and writing thoughts, stories or poetry.

Writing activities also appeal to the interpersonal intelligence. These learners respond best through person to person relationships and communication. These learners also enjoy conversation, listening, being acknowledged and appreciated. The intrapersonal relates to self-reflection, awareness of area, day dreaming awakened by mediation, individuality, and self confidence.

Writing has many benefits, but one of the greatest concerns of teachers is finding the time and how to get students to write. Burns suggests different ways to incorporate writing into the classroom. "Journals or logs help students keep ongoing records of what they do in math class. When students begin to write in their logs, give them general reminders, such as write about what they did, what they learned, and what questions they have. Write about what was easy and what was difficult for them to solve. When writing the solutions to math problems, students can explain their thought processes. Time to time ask students to write about a mathematical concept" (Burns,1995).

"Researchers have successfully used journal writing (Nahrgang and Petersen 1986; Borasi and Rose 1989) to get students to express their anxieties about mathematics and the problems they encounter in the learning process. Bell and Bell (1985) found expository writing to be an effective and practical tool for teaching problem solving (Miller,1991).

Students are more willing to write about what they don't understand than to say so in front of their peers. Public exposure is one of the greatest factors of anxiety. Honest feedback from students will enable teachers to look more closely at the way a concept was presented and what ways they can improve their lesson. Portfolios, a collection of students work, can also be used to show the progress students have made over a period of time. Portfolios can also be used as a means of alternative assessment.

HOW TO STUDY MATHEMATICS

Many students have poor study skills in many subjects but it is especially true in mathematics. Many students believe that they can't study for a math test. There are books, pamphlets and videos on how to study math. There is now "computer on line math help" available to students with homework questions.

"It takes skill-not just ability - to study mathematics" (Margenau and Sentlowitz,1997). Students might find it helpful to get acquainted with someone in their class who already is successful in math. Decisions about time and place when planning to study mathematics must also be considered. Students need to keep their work organized and prepared. The greatest enemy of study is the tendency to delay or postpone. Students should come to class prepared as if their teacher is giving a pop quiz (Margenau and Sentlowitz,1997).

Other helpful tips for studying are look through the material to get a general idea of the topic, note especially any new terms or symbols and their explanations, read the instructions and explanations before working any problem, concentrate on why and how of problem solving, follow the order or sequence of steps in the

examples, consider why the steps were done, use practice problems to check whether students have learned the why and the how and use mistakes made to see how the mistake was made and steps necessary to correct it. Students must continue to practice what skills have been learned into everyday life activities, such as purchasing, banking or working (Margenau and Sentlowitz,1977).
Homework - "If you have saved all your assignments, correctly written the details of every problem, time spent reviewing can cut homework time considerably" (Margenau and Sentlowitz, 1977).

This author believes immediate feedback of homework problems is extremely important. This author goes to each student during class and checks for completeness of homework. Credit is given for homework. After everyone has been checked, the answers are given on the board. Students are encouraged to ask questions and give their solutions. Students use this time to make corrections and get immediate feedback from teacher and input from other students. This encourages dialogue between student and teacher.

"Students must remember the main purposes of homework are (1) to help reinforce what was learned in class, (2) help identify those parts of the lesson that was not understood, (3) to increase the number of correctly solved problems to which students can refer when reviewing for a test" (Margenau and Sentlowitz).

TEST TAKING TIPS

Greatest amount of anxiety occurs during test taking time. This is a time often when a certain amount of math problems must be completed within a given period of allotted time. This author has students to turn in test papers at the same time. Along with the test papers, students are given a challenge (enrichment) sheet. Therefore, students who finish early can go immediately to the challenge enrichment activity, while the slower students can continue working on the test. By giving enrichment at the same time the slower students do not know when others have finished. The more advanced students are given a challenging assignment to encourage further study. The greatest amount of anxiety during a test is when students see other students get up and turn their work in very quickly. The slower student then begins to panic and doubt his own abilities. The student would then begin to wonder why hasn't he or she finished already and what is he or she doing wrong. The faster student tends to make careless mistakes because she or he wants to be one of the first to finish. Therefore, it is more beneficial to all students to turn in test papers at the same time.

There are numerous books, magazine articles, classes and computer information to help improve test taking skills. Here are a few tips: (1) Before a student begins any problems, read the whole test. (2) Do the easiest problems first, skip hard ones and go back to them later. (3) Read the problem and plan how to do it before starting to solve it. (4) Solve the problem before looking at the multiple choice answers. (5) Use the answer sheet correctly: Mark answers in the right place. Fill in the circles darkly and completely. Mark only one answer for each problem. Erase stray marks on the answer sheet. (6) Answer every problem; there is usually no penalty for guessing. (7) Eliminate obviously wrong answers. (8) Ask is the answer reasonable. (9) Stay calm and work steadily; do not daydream. (10) Have a positive attitude; if a student believes she can pass, she improves her chance of success. (11) Students should use scratch paper to write down the numbers they need to solve a problem. (12) Students should look for key words that tell them what kind of computation is needed, for example: less than, greatest, between, nearest, least, closest, and so on. (13) Students should make sure they know what to solve for in each problem. Students should write a number sentence or an equation to help them solve the problem. (14) Write down each piece of information given in a problem, and write down or circle what each problem asks. Then after solving the problem go back and make sure it answers the question, circled earlier. (15) Rename fractions with different denominators as like fractions. (16) Students should always reduce fractions to lowest terms. When students look for the correct answer to a problem with fractions, look for the one that has been reduced. (17) When a student is solving a measurement or geometry problem, first write down the formula they need to solve the problem. (18) Students must remember that equations must stay balanced. Whatever operation is done to one side of the equation must be done to the other side. (19) Check subtraction problems by adding; check division problems by multiplying. (20) Students should write clearly each step of the solution. Be neat and don't rush writing numbers down. Neatness makes it easier to recheck work. (21) Double check their calculator work immediately, use their calculator twice on each problem. (22) Work at a moderate steady pace. Don't spend too much time on a problem; go on to a easier one then come back later to the difficult one. This helps keep frustration level down.

Students that continue to study and practice the math tips provided will have greater confidence in their work and less math anxiety. They will go into a test and solve math problems. This systematic approach will empower students to believe they control the math problems and not math problems control them. This approach allows students to focus on the subject of mathematics rather than on their fears of mathematics.

FAMILIES AND MOTIVATION

Family is one of the four major influences on a childs' motivation to learn according to Wlodkowski and Jaynes, 1990. "For more than twenty years, research has shown that students from elementary level through high school benefits from family conditions and practices that emphasize and encourage learning in school"(Wlodkowski and Jaynes,1990). "The research of Benjamin Bloom supplies vivid evidence for this kind of impact. His research team conducted in-depth interviews with talented and very successful young professionals (ages twenty-eight to thirty-five) who are highly recognized in difficult, competitive fields, including research mathematics, neurology, classical piano, and tennis. He found the most common characteristic of their general education, specialized training and subsequent achievement was enthusiastic parent involvement. Even when the accomplishments and expertise of these young people excluded their parents' direct involvement, they saw their parents' support as the main reference point reinforcing their goals as worthwhile and within their reach" (Wlodkowski and Jaynes,1990,p.15).

Regardless of family income, education and ethnic background, children of families that displayed a number of positive attitudes and high expectations of their children, will be more motivated and active learners.

"When parents understand, value, and nurture the healthy struggle of mathematics, parents can become a strong ally and assist teachers" (Sutton,1997,p.49). Sutton suggest "Let parents know the struggle is okay, that it takes time to learn things, and multiple passes through the same material are often necessary." With this message and this acceptance coming from parents, students can relax enough to learn new concepts. If parents believe that struggling means stupidity or shirking, students feel a tension that gets in the way of learning. Parent's acceptance is very important.

Sutton also suggests: parents should resist the very common temptation to explain the struggle as a in - born difficulty in mathematics, genetic or otherwise. Parent should not say "I was never very good at math, either." This has been stated to this author many times over the years when others have been told the author is a mathematics teacher. This has been stated even in parent conferences. Without realizing it, parents are telling their children it's okay, it's a family tradition and why the child is not doing well in math. Parents need to encourage their children to work hard through to understanding, to dig in when it gets tough, not flee into excuses.

Student's textbooks can be a great assistance to parents, without needing to understand the math involved. When parents see what good resources textbooks can be, they begin to see how to help their children use these books.

FAMILY MATH is a wonderful new development in mathematics education. FAMILY MATH means parents and their children doing mathematics together and having fun doing it. The purpose of the program is to create a friendly environment where parents and other caregivers work together with children on various activities designed to include and supplement the school math curriculum. The activities involve doing (hands on math) and help children to see how math and daily life connect (Zaslavsky, 1994).

The activities include making a survey of the group to find out each person's favorite ice cream flavor, then make graphs to illustrate their findings. Activities may be to make decorative designs and discuss the symmetry involved. The group may analyze the strategies involved in the games they play. They even create their own puzzles and games. While having fun in a cooperative atmosphere, they deal with many mathematical topics - arithmetic, geometry, measurement, probability, statistics, graphing, calculators and computers and logical thinking.

The activities require inexpensive materials that participants can take home for further practice. A FAMILY MATH book is available. The sessions are headed by classroom teachers, parents and or community leaders. Speakers are often invited to discuss their careers and opportunities in the field of math and science, to encourage children to think about math - related careers (Zaslavsky, 1995). Teachers have seen the benefits of the program in their daytime classes. Students have seen that math can be fun. Parents are left with more confidence in assisting their children in learning.

Xavier University, a historically black institution in New Orleans, runs summer enrichment programs for secondary - level students requires parental involvement. Parents are required to attend an orientation session. They receive report cards frequently and attend an awards assembly at the end of the program. Students receive a certificate and have opportunity to demonstrate what they have learned (Zasklavsky,1994).

The National Council of Teachers of Mathematics has published a brochure entitled, "Help Your Child Learn Math". Suggestions made in the brochure were the following: (1) Ask your child's teacher about the kinds of help that you as a parent can provide. (2) Encourage their child to putting the word problem or definition in the child's own words. (3) Parents should make sure that "home" math has a noticeable problem solving flavor. It should contain a challenge or questions that must be answered ("How many nickels do you have in your bank?", "How many do you need to buy an ice cream cone?" (4) Parents should use objects that children can touch such as soft toys, blocks, marbles, drinking straws and fruit to teach a concept. The National Council of Teachers of Mathematics brochure also suggests parents reward their children with praise for the correct answers and effort made. This builds the child's confidence in problem solving.

Parents must remember to have patience with their children. If a parent encounters a skill their child cannot seem to comprehend, try another technique, try another strategy, or just wait for another day. "The key to teaching is simple: the more positive the climate remains during instruction, the more receptive a child is to new ideas" (Yablun,1995).

Parents have a greater opportunity to show their children the everyday uses of mathematics such as sale items, balancing a checkbook, measuring while cooking, gardening, etc. Many of these activities involve hands on activities and one-on-one learning situations. One-on-one situations are less intimidating to children than the regular classroom setting. This one-on-one interaction allows children to feel more comfortable about asking questions.

Parents want to help their children and see them successful in mathematics. Teachers must make parents feel welcomed and appreciated at the very beginning of each school year. This author gives each parent the poem entitled "Sculptors" along with class expectations at the first day of class. Teachers must encourage parent input. During an open house session, addressing a group of parents, this author reminds them that one plus one is still two and that Algebra is only math with alphabets thrown in. This author reassures parents to show their children a concept the way the parent remembers learning it in school. Many parents feel that math has changed since they were in school. Yes, many methods have changed, but the basic concepts are the same. This author encourages students to learn different ways to solve a problem and cheer when they share different methods of problem solving with the class.

MATH PROJECTS

Math projects and fairs are a great way to get parental involvement. Most of the students' work is done outside of classroom. These projects and fair competition would give parents an opportunity to have an active role in the design, research and organization of the math project.

The project method of teaching was first introduced more than fifty years ago by John F. Woodhill of Teachers College, Columbia University. The fairs were first used by art, home economics, industrial arts and vocational agriculture then science. Now mathematics fairs for students have been conducted at all levels, beginning with the first grade (Hess, Allinger and Andersen,1992).

"The use of technology (calculators, computers, videodisks, etc) in the mathematics classroom offer wonderful opportunities for student projects" (Hess, Allinger and Andersen,1992).

A mathematics project consists of all the effort expended in solving a problem, exploring an idea or applying a mathematical principle that is, the initial planning, the study, the exhibit and written report. The exhibit uses drawings, graphs, models, pictures, words to briefly tell the viewer the students' idea of the mathematical concept or principle, show the use and answer a question posed by the author (Hess, Allinger and Andersen,1992).

A written record should be kept at every stage of the project. Students will broaden their backgrounds in mathematics and explore mathematical topics, they never knew existed. In setting up exhibits, they will experience the satisfaction of demonstrating what they have accomplished. Students will grow in self-confidence as they share their experiences with others (Hess, Allinger and Andersen,1992).

There are many books and articles available on selecting math projects. Here are a few ideas: (1) Design a Park (2) Plan a Talent Show (3) Design a Theater (4) Construct Geometric Solids (5) Conduct Survey Research (6) Observe Traffic Flow (7) Gather Environmental Data.

Students can also do research on the biographies of famous mathematicians. The author believes children need to see mathematicians as ordinary people. Hopefully with this insight children will begin to understand that mathematics is a subject that can be done by all not just a certain few.

The author finds the biography of Albert Einstein an inspirational story that is shared with students. Albert Einstein was a German-American physicist and mathematician, whose major contribution was the Theory of Relativity. During childhood his teachers thought he was stupid and he failed math in elementary school. He did not display his talent for mathematics until he was 14, when he taught himself integral calculus and analytical geometry (World Book Encyclopedia,1994).

Einstein's biography is stressed to students to show them each year is a new year full of hope. Students need to keep an open mind with mathematics and disregard previous failures involving mathematics. Previous negative feeling regarding mathematics is one of the greatest causes of math anxiety.

STUDENTS' INITIAL SURVEY

Teachers must face many concerns and needs of students in the classroom. The author took a survey regarding students feeling toward math the first day of school. Students were also asked about what was their greatest math concerns. The author stated to the students it was the author's hope to make math an enjoyable, fun and relaxing experience as possible this year. Lessons will be taught in a variety of ways and suggestions by students are welcomed.

Students made some of the following comments regarding their feelings toward mathematics:

> "I don't really like math. My reason is it seems to be hard and confusing to me. I always lose concentration while doing math."

> "I don't like math very much because I don't understand a lot of it."

> "I don't like math because it's boring and some stuff you learn you may never use it."

> "I feel good about math, it's easy."

> "I don't like math. It has been my hardest subject."

Forty-four percent of the students surveyed said they enjoyed math. Most of these students who like math felt they have been successful in previous math classes. Many who enjoyed math saw it as a necessary part of life and it was a needed skill. Many of the students said they enjoyed working numbers and saw math as a challenge, fun and/ or interesting.

Forty percent of the students said they did not like math. Many of these students based this on their previous math experience. Some students stated math was their worst subject and that they've never done well in math. Most of these students said they found math confusing and/or boring. A few believed many concepts taught in a math class would not be used later in life.

Sixteen percent of the students surveyed said they had mixed feelings regarding math. Many said it depended on what concept they were working on. Most said it depended on how the teacher taught math and rather or not it's made fun and explained well.

Students were also asked which ways do they prefer to learn. A formal learning style survey will be given later. Forty-one percent of students preferred seeing such as using pictures, writing on the board, videos, overhead usage. Forty-seven percent of students preferred hands on activities. Only twelve percent of students said they preferred hearing as a method of instruction, which is using the traditional lecture by teacher.

It is apparent from the survey students feel differently about mathematics. Many of the students feel they have not been successful in math and see this as the way things will continue. The students also prefer different ways the lessons were taught. The author will teach in a variety of ways and keep in mind the comments and concerns of students.

STUDENT BOREDOM

Boredom was stated in the survey as one of the reasons students disliked math. Students should select topics that are of interest to them when doing math projects, many homework assignments and classwork. Information could be gathered from home or classmates when doing an assignment. For example, if students are working on making circle graphs, let students gather information about themselves. The author has had student's to make a survey about the class' favorite color. This information was gathered, then circle graphs were made. Students found this more enjoyable because it was information about themselves. The numbers had more true meaning than numbers taken from a textbook.

Monotony is frequently a cause of boredom. Doing the same thing repeatedly without any real change becomes dull. Learning, with its demands for practice and routine, can become quickly tedious to many students. Constraint contributes to boredom as well. When people feel confided, the tasks they are doing become more oppressive to them. When students do not value the purpose to their assignments, they will see the work as wearisome (Wlodkowski and Jaynes,1990).

A task that is wearisome and boring will cause a great deal of anxiety. A class that is boring will cause students to experience anxiety as soon as they walk into the door before class even begins.

"If the student is willing to put forth the necessary effort, the teacher has a responsibility to help that child feel successful" (Wlodkowski and Jaynes,1990).

Wlodkowski and Jaynes (1990) suggest some methods that make learning more stimulating such as: (1) provide variety in learning, (2) relate learning to student interests, (3) use unpredictability within safe bounds for learning, (4) use novel and unusual teaching methods and content with students, (5) give students questions and tasks that get them thinking beyond rote memory, (6) have students actively participate in learning, (7) provide consist feedback, (8) create learning experiences that have natural consequences or finished products, (9) use cooperative learning techniques, (10) encourage student choice in learning situations, and (11) offer learning that is challenging.

Students will let teachers know what activities they really enjoy. They will show the teacher while the activity is going with such indicators as less interruptions, and more student participation. Students will also ask when a class activity can be done again.

THE NEED FOR MANIPULATIVES

"Children learn best when they are active rather than passive learners" (Spikell,1993,p.4). According to Spikell (1993) most learners, whether adults or children, will master mathematical concepts and skills more readily, if they are presented first in concrete, pictorial and symbols. Manipulatives are concrete objects used to teach a concept. By using manipulatives, pictures and symbols to model or represent abstract ideas, the stage is set for learners to understand the abstractions they represent.

Manipulatives appeals to the learning style of kinesthetic learners because they are able to actually touch the objects. Manipulatives are important in the teaching and understanding of mathematics. "Almost every mathematics idea, except simple arithmetic facts, consists of three components: linguistic, conceptual and skill/procedural" (Sharma,1987). The conceptual component deals with the center of mathematical concept. The child needs a model to conceptualize the idea (Sharma,1987).

Sharma (1997) feels there are six levels of the mastery of mathematical concept: intuitive, concrete, representation (pictorial), abstract, applications and communication. "Ideally, each mathematics concept should be introduced beginning at the communication level. However almost all mathematics teaching activities take place at the abstract level" (Sharma,1997).

Students have a tendency to forget when taught only at the abstract stage. Thus students become frustrated because mastery was never fully attained. Students will begin to have difficulty in learning mathematics. The results in failure will cause many students to develop a fear of mathematics (Sharma,1997).

Pictures can also be used to present a concept appeals to visual/spatial learners. "Visualization is the natural way one begins to think. Before words, images emerge" (Sharma,1983).

Manipulatives helps relieve boredom in children allowing them to explore and use their imagination. The manipulatives provides a picture of a math concept. Manipulatives offers a change from the textbook (abstract) method of teaching. Manipulatives also can be placed within cooperative groups, which is appealing to the interpersonal learners. These are some of the benefits of using manipulatives in the classroom.

Many manipulatives are inexpensive and can be everyday objects. Money, 2 - color counters, calculators, rulers, dominoes, playing cards, buttons, and number cubes are a few of the commonly available manipulatives that can successfully be used in the classroom. These manipulatives can be used to teach such concepts such as angles, area, decimals, factoring, estimation, fractions, measurement, counting, percent, prime numbers, probability, geometry and whole numbers.

There are companies which specialize in manipulatives that can be ordered from a catalog. No matter where a school is located, materials can be made available through the mail. These companies also make available to teachers such manipulatives as tangrams, pattern blocks, geoboards, algebra tiles, cuisenaire rods, miras and polyhedral models.

Many teachers feel as though they do not know how to teach using manipulatives and, therefore, hesitate to use them in the classroom. Many math teachers, who attended college more than ten years ago, were usually taught on the abstract level, textbook, pencil and paper, all through their school days. There are classes and workshops for teachers to learn how to teach using manipulatives. The companies that make the manipulatives also provides books and pamphlets on ways the material can be used. There are articles on using manipulatives in mathematics teaching journal such as the NCTM's Arithmetic Teacher magazine.

Manipulatives require a great deal of prior planning and organization. This educator uses sandwich bags to keep many of the manipulatives organized and ready for student use. Students enjoy the change from lecture and books. They are more inclined to explore with manipulatives and show greater interest in classwork.

GAMES ENHANCEMENT OF MATH SKILLS

Games with manipulatives may allow students to apply what they learned to the real world. Using board and card games along with cooperative learning are ways that students can become involved in a positive mathematical environment. Games are highly motivational to students and can be used effectively to practice specific skills. Games can also be used to introduce children to other cultures.

"Using games in the classroom and at home will maximize students' problem-solving competence, ability to communicate and reason mathematically, perception of the value of mathematics, and self-confidence in their ability to apply mathematical knowledge to new situations"(Leonard and Tracy,1993).

COOPERATIVE GROUPS IN THE CLASSROOM

Small group cooperative learning provides an alternative to both traditional whole class instruction and individual instruction. "Systematic and frequent use of small group procedures has a profound positive impact upon the classroom climate; the classroom becomes a community of learners, actively working together in small groups to enhance each person's mathematical knowledge, proficiency, and enjoyment" (Davidson,1990).

Human beings have strong needs for contact and communication with others. This is especially important to school age children. Many students are motivated to come to school in order to be with their friends; they have a strong need to be accepted, to belong and sometimes to influence others (Glasser,1986).

Cooperative learning makes use of basic characteristics of human nature. Working in groups rather than isolation helps relieve math anxiety. "The learning of mathematics is often viewed as an isolated, individualistic matter. One sits alone with paper, pencil and perhaps calculator or computer and struggles to understand the material or solve the assigned problems. This process can often be lonely and frustrating. Perhaps it is not surprising that many students and adults are afraid of mathematics. They are troubled by math avoidance or math anxiety" (Davidson,1990,p.3).

Cooperative groups provide students a chance to exchange ideas, to ask questions freely, to explain to one another, to clarify ideas and concepts, to help one another understand the ideas in meaningful ways and express feelings about their learning. This interaction meets the needs of the verbal/linguistic, intrapersonal and interpersonal learners. Cooperatives group learning offers opportunities for success for all students in mathematics. Students within groups are not competing one against another but are working together (Davidson,1990).

The role of small groups in mathematical communication is addressed in the Curriculum and Evaluation Standards for School Mathematics by the National Council of Teachers of Mathematics (1989): small groups provide a forum for asking questions, discussing ideas, making mistakes, learning to listen to others' ideas, offering constructive criticism, and summarizing discoveries in writing. Cooperative groups maximizes the active participation of each student and reduces the isolation of individuals. Isolation is a major contributing cause of math anxiety.

The negative effects of isolation and the benefits of cooperative learning is recalled by the author. While the author was explaining how to solve equations, a student looked confused. He said he didn't understand. The author explained again but he still seemed confused. He said he understood but his facial expression said differently. To relieve him from feeling isolated and singled out, the author went on with the lesson.

However, during the next class meeting everyone sat in groups of four. Questions were placed on the overhead. A part of the solution was given. Students had to discuss within their groups and give the next step on a group flash card. This process allowed equations to be solved at a slower pace. Students who might hesitate asking questions in front of the entire class now had an opportunity to ask their classmates within the small groups.

The same student, mentioned earlier was answering questions with his group and asked for the next question. He seemed relaxed and happy along with many of the other students. After groups made their decision, group flash cards were held up at the same time. The groups earned team points. The students were laughing, eager for the next question. Students made the following comments such as, "This was fun. The class period sure did go by fast. I hope we can do this again." Students were learning and having fun at the same time.

COMPUTER ASSISTANCE

Of course there will be times when math will have to be studied alone. This is the time when a lot of students feel frustration while studying or doing homework alone. Technology has provided assistance at home through the Internet and math computer software now available. The Internet service called Dr. Math, helps students with math problems. In the fall of 1994 the Math Forum of Swarthmore College started a project where K - 12 students can send in math questions and get personal answers. The service grew rapidly in three years. By 1997, there had been over 150 volunteer 'Doctors' from all around the world (Forum, Swarthmore,1997).

Students submit questions to Dr. Math by filling out web form or by sending in e-mail. Answers are sent back by e-mail. These questions and answers are gathered into a searchable archive organized by grade level and topic. The information then can be used by others (Forum, Swathmore,1997).

There are many homework services available on the Internet. Through the Internet, students do not feel isolated, a contributing factor of math anxiety. Students also will feel less hesitant to ask questions because the fear of being discovered not knowing, has been removed. Students know their identity is unknown.

Computer use at home and in the classroom has increased greatly over the years. There are many software programs to make math fun. Algebra Blaster is a very popular software program. In the program students go on an outer space adventure and solve various problems to return home. The computer gives students immediate feedback and is done at a fast pace. This is the way children are used to playing different arcade games.

Children are comfortable with computers and enjoy them.

There are also software and equipment that can be used by teachers to project what appears on a computer monitor to a large screen. The entire class will be able to view at the same time and discuss. Books are available such as Using the Internet to Investigate Math. Many math websites for student and teacher usage are accessible. This is another way to help math come alive and relieve boredom. Boredom is one of the greatest complaint about math from students.

MATH EFFECTS ON SOCIETY

It is important that math is seen as fun and exciting. The ill feeling toward math has long term effects such as math avoidance. Math avoidance causes many people to avoid certain careers.

The economy of the United States has changed greatly within the past few decades. There has been great growth in communications and finance, while manufacturing jobs have gone abroad or have been restructured to require higher-level skills on the part of the workers. According to the Bureau of Labor Statistics 60% of the jobs were unskilled, 20% skilled and 20% professional in 1950. In 1991 the number of unskilled jobs declined to 35% and skilled jobs increased to 45% and professional jobs remained the same at 20%. For the year 2000 unskilled jobs would account for only 15% of the workforce and skilled jobs would for 65% and professional jobs skills would remain the same at 20%. With advances in technology occurring frequently, people will have to learn to work smarter, rather than faster as in the past. More jobs will require a knowledge of mathematics. New technologies will call for the ability to apply mathematics and science in practical ways, and rapid changes will require workers to learn new skills throughout their lives (Zaslavsky,1994).

Entry tests and civil service exams for many jobs usually include questions involving mathematics. The greatest growth industry predicted is information processing, and computers will play a major role. The computer is a part of many fields such as health care, music, theater, fine arts and many others (Zaslavsky,1994).

Many people are not prepared for the great demands of society. "Public acceptance of deficient standards contributes significantly to poor performance in mathematics" (National Research Council Report Everybody Counts,1989). People willingly say, "I am no good in math". But how many people will admit they can not read? Society accepts math incompetence as the norm.

"Three of every four Americans stop studying mathematics before complete career of job prerequisites" (National Council Report Everybody Counts,1989). As a result many companies, universities and armed services are forced to provide extensive remedial education.

There is a supply and demand problem for many jobs remaining in the United States. Around the country, a shortage of skilled labor dilemma: Thousands of people are being turned down for factory work by companies that are actively recruiting. The result: unfilled jobs and very frustrated job seekers (Narisetti,1995).

Today's factory workers need to use advanced math, have good communication skills and understand the use of computers (Narisetti, 1995). Most people, who are returning to school due to career changes, say they would have taken more math if they knew then what they know now.

Professor Hassler Whitney stated math avoidance occurs because students are not given enough opportunities to see and do real math (Zaslavsky,1994). This is also a major complaint from students about the math taught in the classroom. They do not see how it affects their world now or in the future.

Careers are presented in most math textbooks. Many of the careers show the math used for that job based on the math concept being covered in the textbook. Hal Saunders, a math teacher interviewed people representing 100 different occupations to find out which of some 60 math topics they used, and how they used them. As a result of the interviews, he formulated 435 word problems in his book entitled When Are We Ever Gonna Have to Use This? (1988). Saunders' book is organized so a math topic will show how the different careers will use that particular topic on the job. His book is also organized so that someone interested in a certain occupation can find the different math which would be used on their job.

Math anxiety decreases when math is seen as a tool to be used. The earlier people are made aware of the need for mathematics, they will become determine to overcome their math anxiety. Mathematics is a creative, dynamic, growing area of thought and action and it is not at all dry or dreary. If students are shown how useful math is in all subjects it could turn their fear into curiosity.

MATH ANXIETY

Changes in society are causing many older Americans to make career changes. They are changing to careers that require more mathematics than they already have. Therefore many are returning to school and math can no longer be avoided.

The reasons for mathematics avoidance have been mentioned earlier. Many people suffer from math anxiety. Math anxiety has been defined as the "emotions" clutter one's understanding and recall as one attempt to solve math problems. They cause one to forget and lose one's self-confidence.

Over the past two decades there has been a large body of research documenting math anxiety among students from elementary through college years (Kagan,1987). Several researchers have studied the effects of math anxiety, looking for predictors and products (Meece, Wigfield, and Eccles,1990). Hembree (1990) tried a number of different methods to reduce anxiety (Zyland and Lohr,1990). Some of the suggested origins of mathematics anxiety were environmental, intellectual and personality factors. Many feel that the problem is primary environmental (Hartshorn,1982;Kogelman & Warren,1978), while others proclaim that it is more innate and is a function of cognitive style or poor mathematics talent (Betz,1978; Frary & Ling,1983; Hadfield of Maddux,1988).

Here are a few symptoms of math anxiety: (1) panic sets in, (2) intense pressure to find the correct answer, (3) tension builds from time pressure, (4) uncertainty develops from a lack of self-confidence, (5) forgetfulness set in, (6) doubt in one's intelligence to do the job, (70) can't think, and (8) pencil stops moving.

Hembree found significant math anxiety reduction through systematic desensitization. System desensitization is a technique that combines muscle relaxation with imagining exercises. The treatment is self-administered by the student. The taped program could be used in the privacy of his or her home during convenient times (Zyl and Lohr,1990).

Tobias (1987) suggests to reduce math anxiety, one must learn to first to recognize when panic starts, then to identify the static in one's analytic and retrieval systems and finally clear up the static without ceasing to work on the problem as one works on oneself. "The essence of math-anxiety therapy is self-monitoring. Self-monitoring can be done while working on homework or studying" (Tobias,1987). Tobias suggests the divided-page exercise to learn to give permission to explore the confusion and to find out what is making the problem or the new material seem difficult. Therefore one is writing and working continuously. Writing things down frees one from the paralyzing effort of staring at a hard problem or a page of confusing text. "Because thinking involves doing, writing down random feelings and thoughts breaks the tension and sense of isolation"(Tobia,1987).

Many colleges and universities offer overcoming math anxiety courses. A course offered at Northeastern University was described as a hands-on workshop. A workshop presenting a comprehensive diverse approach to math anxiety that will help students redirect their thinking in a way that will promote self-confidence and success in mathematics (1997).

Western Michigan University Academic Skills Center provides the following suggestions to help overcome math anxiety: (1) Begin in an appropriate level class. Don't take advanced classes without the proper background. (2) Set aside specific time every day for studying math; just going to class is not enough. (3) Get help when you need it. Ask an instructor, friends or get a tutor.

(4) Face anxiety; feel it for few moments, then practice relaxation techniques such as deep breathing to help calm down physically. (5) Don't allow What if...? questions to take over. Instead say "I can" or "I want to." This leads to positive problem-solving approach. (6) Focus attention away from oneself and back on the problem at hand. (7) Practice problems until one is comfortable with them.

Dr. Dorothy Buerk, a mathematics professor at Ithaca College suggests the following principles to relieve math anxiety: (1) Experience the problem, relate it to one's personal world, clarify the language so that it makes sense (2) Rely on one's intuition and feelings (3) Look at the limitations of any solution to a problem and the conflicts that remain (4) Make exceptions to the rules when one believes it is appropriate (5) Be reluctant to judge (Zaslavsky,1994).

Another approach to overcoming math anxiety is the clinical approach. In this approach students are given assertiveness training. "This technique, designed by feminist therapists, is meant to give students the training they need to survive in the next class. The Math Anxiety Bill of Rights was developed by Sandra L. Davis to help students articulate typical concerns about math (Tobias,1997).

Math anxiety is very real and occurs among thousands of people. Much of this happens in the classroom due to the lack of consideration of the different learning styles of students.

LEARNING STYLE & FOLLOW UP STUDENT SURVEY

"Students often feel lost and powerless because of the mismatch between the style of teaching they encounter in the classroom and their own personal learning style" (Zaslavsky,1994). There are many factors to be considered when analyzing personal styles in learning mathematics. Some of them include (1) preference for working alone or working in a group, (2) preference for a competitive or a cooperative environment, (3) preference for oral or for written methods, (4) manner of tackling a new situation or problem, (5) reflective style or an impulsive style, (6) speed of work, (7) persistence in staying with a task, (8) need for outside encouragement and reinforcement, or a go-it-alone attitude, (9) willingness to take risks, and (10) spatial, numerical, logical, kinesthetic, or other approaches to problem solving according to Zaslavsky (1994). After considering these factors it is apparent everyone thinks and learns differently. These differences must be considered in the classroom.

The author's students were given a learning style survey (Appendix, p. 84) two months after the first survey was given. The results of the survey showed that most of the students had interpersonal learning style preference. The other learning styles most selected by students were intrapersonal, musical, visual then body/kinesthetic. The least preferred learning styles by students were verbal then logical. The two learning styles least preferred by students are the two that are generally taught to (Cassone and Cassone,1991).

Therefore there is generally a great mismatch which leaves students feeling lost and powerless causing math anxiety (Zaslavsky,1994).

The author taught students in a variety of ways over a two month period. Class activities included use of cooperative groups, art work, manipulatives along with textbooks and lectures. Fifty-six percent of the students surveyed said they felt better about math. Forty-four percent of the students surveyed said they felt the same or worst.

Students who felt better about math, said math was more fun and interesting. Many of the students believed math was easier and they were learning more. Students comments about their feeling toward math changed for the better because of some of the following reasons: (1) I am doing different things. (2) I am passing. (3) I am finding out math can be fun. (4) I enjoy math class now because I finally have a teacher to really explain things. (5) It's getting more interesting. (6) The teacher made it fun to learn. (7) I like it more because we do lots of fun things.

CONCLUSION

Negative feelings toward mathematics is very common, upon people of various backgrounds, race and gender. These negative feelings affects student achievement and self-esteem. People have math anxiety because of these negative feelings which remains for many throughout the rest of their lives. The author felt that if different approaches to teaching math are provided, students' achievement and confidence would improve. Therefore their feelings toward math would become positive and their anxiety reduced.

Students were surveyed the first day of class. Their feelings toward math were based on prior math experiences. Many students felt they were not good in math and math was useless, confusing and boring.

Over the next two months, students were taught using various methods of instruction. Math instruction would include lessons using increased use of manipulatives, open-ended questions, alternative assessment and cooperative groups.

Students' learning styles would later be examined. Another survey would be done regarding students' feelings toward math. Most of the author's students had interpersonal learning style preference followed by intrapersonal, musical, visual then body/kinesthetic. The least preferred learning styles were verbal then logical. The two learning styles least preferred are the two most commonly taught to.

Clearly as a result of this great mismatch students often feel frustrated, confused and defeated. The impact is so great that later in life, many adults will avoid math even in making career choices. This avoidance of math will affect their livelihood and self-esteem.

This society no longer has a great need for factory workers, but instead has a greater need for technical workers, with a strong math background. Employers have expressed the frustration of not being able to find highly skilled workers.

The needs of society requires a greater need for mathematics. Math must be looked upon in a positive light to reduce math anxiety. Therefore teachers must re-examine traditional teaching methods, which often do not match students' learning styles and skills needed in society. All students do not learn the same way all the time. Lessons must be presented in a variety of ways. The different ways to teach a new concept can be through play acting, cooperative groups, visual aids, hands on activities and technology.

Math reforms call for students, who are able to understand mathematics, can think and reason mathematically and use what they've learned to solve problems both in and out of school.

Math anxiety is very real and affects many people. Math anxiety can be overcome through various methods. Teachers are responsible to help their students be more successful. Teachers need to help students see the usefulness and the excitement of learning math. Once students see mathematics as a tool to be used rather than an obstacle, they will become determined to overcome their math anxiety.

As a result of the change in students' attitudes toward mathematics in a short period, the author hopes to continue doing research on math anxiety and find other creative ways to make learning mathematics enjoyable to students. Other teachers have noticed the author higher student success rate (less failures) at the end of the first nine weeks marking period. The author also has had less disturbances and fewer discipline problems than many. Students are more engaged and now see math as fun and they feel they are learning more.

REFERENCE LIST

About Ask Dr. Math. (1997, October 10, 8:23).On-Line].available:http://forum.swarthmore. edu/ dr.math/abt.drmath.html

Bandalos,D., Yates, K.,& Thorndike-Christ, T. (1995). Effects of math self-concept, perceived self-efficacy, and attributions for failure and success on test anxiety. Journal of Educational Psychology,87 (4) 611-623.

Board on Mathematical Sciences and Mathematical Sciences Education Board, National Research Council. (1989). Everybody counts: a report to the nation on the future of mathematics education. Washington, D.C.: National Academy Press.

Burns, M. (1992). About teaching mathematics: a k-8 resource. White Plains, N.Y. Math Solutions Publications.

Burns, M. (1995,April). Writing in math class? Absolutely! Instructor, pp. 40 - 47.

Buxton, L. (1991). Math panic. Heinemann Portsmouth, N.H.: Dell.

Cassone, P., & Cassone, L. (Eds.). (1990). Accelerative learning and teaching. Minneapolis, M.N.: Internal Center for Accelerative Learning.

Course descriptions: mathematics, (1997,October 9, 7:52). [On-Line]. Available http:// www.northeastern.edu/uc/catalog/MTH.ht

Dauben, J. (1994). Mathematics World Book Encyclopedia, 302-308. Chicago, I.L.: Field Enterprises, Inc.

Davidson, N. (Ed.). (1990). Cooperative learning in mathematics: a handbook for teachers. New York: Addison-Wesley Publishing Company.

Davis, S. (1991,Summer). Math anxiety bill of rights. College Teaching, 39, 91-93.

Dubbey, J. (1970). Development of modern mathematics. New York: Crane, Russak & Company, Inc.

Forte, I., & Schurr, S. (1996). Integrating instruction in math. Nashville, T.N.: Incentive Publications, Inc.

Frank, M. (1986). Easy to make and use: Math bullentin boards. Nashville, T.N.: Incentive Publications, Inc.

Guidelines for dealing with text anxiety. (1997, October 9, 8:02) [On-Line]. http://www. stthomas.edu/www/lab_http/sgs/TstTak7.htm

Guidelines for math exams. (1997,October 9, 8:02) [On-Line]. Available: http:// www.stthomas.edu/www/lab_http/sgs/TstTak8.htm

Hess,A., & Allinger, G. (1992). Mathematics projects handbook, Reston, V.A.: National Council of Teachers of Mathematics

Hirsch, E.D., Jr. (Ed.). (1992). What your 4th grader needs to know, New York: Doubleday.

Independent Study High School Courses - Mathematics -GIBM3 Overcoming Math Anxiety (1997,October 9: 7:54). [On-Line]. Available: http:// www.extend.indiana.edu/hs/deptMath.htm

Leonard,L.,& Tracy,D. (1993,May). Using games to meet the standards for middle school students. Arithmetic Teacher, p. 499-503.

Margenau, J., & Sentlowitz, M. (1977). How to study mathematics. Reston, V.A.: The National Council of Teachers of Mathematics.

Motz,L. & Weaver, J. (1991). Conquering mathematics: From arithmetic to calculus. New York: Plenum Press.

Narisetti, R. (1995, October 12). Worker shortage worries factories. Wallstreet Journal. p. 1-C, 3-C.

National Council of Teachers of Mathematics (1990). Help your child learn math. Reston, V. A.: National Council of Teachers of Mathematics.

Overcoming math anxiety. (1997, October 9:9:53). [On-Line]. Available: http://www.tacoma.ctc.edu/catalog/courses/hum_devl.htm.

Overcoming math anxiety. (1997, October 7:8:15). [On-Line]. Available:http://www.w.mich.edu/asc/math-anxiety.htm.

Paulos, J. (1988).Innumeracy: mathematical illiteracy and its consequences. New York: Hill and Wang.

Saunders, H. (1988). When are we ever gonna have to use this? Palo Alto, C.A.: Dale Seymour Publications.

Sharma, M. (1987). How to take a child from concrete to abstract. Math Notebook, 5, 8-10.

Sharma, M. (1997, July). Improving mathematics instruction for all. Fourth Lecture in the Colloquium Series "Improving Schools From Within: Your Role. pp. 2 - 12.

Spikell, M. (1993). Teaching mathematics with manipulatives: A resource of activities for the k-12 teacher. New York: Allyn and Bacon.

Sumter County(South Carolina) Development Board (1996, November). Employer Expectations Matrix

Sutton, S. (1997, February). Finding the glory in the struggle: helping our students thrive when math gets tough. Math Instruction p. 43 - 52.

Tobias, S. (1987). Succeed with math: every student's guide to conquering math anxiety. New York: College Entrance Examination Board.

Tobias, S. (1993). Overcoming math anxiety. New York: W. W. Norton & Company.

Weissman, M., & Monse, K. (1992). Professor Weissman's Laugh with math. Lafayette, N. J.: Laugh & Learn.

Worthy, Leon, Cambridge College, Cambridge, MA,(1997) Learning Style Survey

Wlodkowski, R., & Jaynes, J. (1990). Eager to learn: helping children become motivated and love learning. San Francisco, C. A.: Jossey-Bass Inc.

Zaslavsky, C. (1994). Fear of math: how to get over it and get on with your life. New Burnswick, N.J.: Rutgers University Press.

APPENDICES

INTELLIGENCE TO INSTRUCTION

INTERPERSONAL:
-operates best through person to person relationships and communication
-promoted by communication, conversation, listening, being acknowledge, and appreciated
-relies on all of the product possibilities of other intelligences

INTRAPERSONAL:
-relates to self reflection, awareness of area, day dreaming, awakened by mediation, individuality, self confidence.

PRODUCT POSSIBILITIES: individual products, independent studies, development of hobbies, science fiction, creative approaches work best, hands on activities, technological challenges

VERBAL/ LINGUISTIC:
-awakened by hearing, speaking, reading, writing thoughts, stories or poetry

PRODUCT POSSIBILITIES: ballad, commercial, rap, announcement, story telling, free verse, jingle, characterization, comparison, riddle, mock interview, debate, oral report, comedy act, commentary, advertisement, campaign speech

LOGICAL/ MATHEMATICS:

-"scientific thinking", promotes reasoning, numbers and the recognition of abstract patterns -
-does well with cause and effect

PRODUCT POSSIBILITIES: puzzles, creating patterns, chain of events, graphs, charts, bulletin boards, predictions, logs, lists, observations, predictions, diagrams

VISUAL/ SPATIAL:

-relies on sight, the ability to visualize and create mental object
-is awakened by vibrant colors, interesting or unusual designs

PRODUCT POSSIBILITIES: albums, scenarios, commercials, timelines, word games/searches, flash card, charts, costumes, travel logs, web, outlines, newspaper advertisements, museums, painting, photography, design

MUSICAL/RHYTHMICAL:

-recognition of a pattern of tones, rhythm and beats
-awakened by hear sounds: environmental, musical
-high level of expression

PRODUCT POSSIBILITIES: writing songs, lyrics, ballads, creating sounds or designs, dance, physical expression, demonstration, charade, dramatization, role play, story telling, rap, rhyme, quatrain, limerick, audio tape recording, recorded music, performing music

BODY/KINESTHETIC:

-physical movement and knowledge of the body
-awakened by walking, dancing, imagining feeling, expressing through physical motion

PRODUCT POSSIBILITIES: charade, demonstration, display, dramatization, experiment, game, mobiles, soap sculpture, model, mime, marionette, invention, etching

LEARNING STYLES SURVEY

____ Has a lot of friends
____ Socializes a lot at school or at home
____ Seems to be "Street Smart"
____ Gets involved in after school activities
____ Serves as the family "mediator" when disputes arrive
____ Enjoys playing group games with other children
_____ Has a lot of empathy for the feelings of other
____ Displays a sense of independence or a strong will
____ Reacts with strong opinions when controversial topics are discussed
____ Seems to live in their own private, inner world
_____ Likes to be alone to pursue personal interest, hobby or project
_____ Seems to have a deep sense of self confidence
_____ Seems to march to the tune of a different drummer
_____ Motivates themselves to do well on independent projects

____ Likes to write
____ Spins tall tales or tells jokes and stories
____ Has a good memory for trivia, names, places or dates
____ Enjoys reading in their spare time
____ Spells work easily and accurately
____ Appreciates tongue twisters and nonsense rhymes
____ Likes doing crosswords, anagrams or scrabble
____ Computes arithmetic quickly in their head
____ Enjoys using computers
____ Asks questions such as "Where does the universe end?"
____ Plays chess, checkers or other strategy games and wins
____ Reasons things out logically and clearly
____ Devises experiments to test out things they do not understand
____ Spends time working on logic puzzles
____ Spends free time in art activities
____ Reports clear visual images when thinking about things
____ Easily reads maps, charts and diagrams
____ Draws accurate representation of people or things
____ Likes to see movies, slides or photographs
____ Enjoys doing jigsaw puzzles or mazes
____ Plays a musical instrument
____ Remembers melodies of songs
____ Tells you when a musical note is off key
____ Says they need music in order to study

____ Collects records or tapes
____ Sings songs to themselves
____ Keeps time rhythmically to music
____ Does well in competitive sports
____ Moves, twitches taps or fidgets while sitting
____ Engages in physical activity such biking, skiing
____ Needs to touch people when they talk to them
____ Enjoys scary amusement rides
____ Demonstrates skill in crafts such as sewing, wood work
____ Cleverly mimics others mannerisms, gestures or behaviors

GAMES THAT ENHANCE MATH SKILLS

IDEAL FOR INTRAPERSONAL, VERBAL, VISUAL AND KINESTHETIC LEARNERS

BACKGAMMON

problem solving, communication, reasoning, patterns, algebra

BATTLESHIP

problem solving, communication, reasoning, patterns, algebra, probability, geometry, measurement

CHECKERS

problem solving, communication, reasoning, patterns, probability

CHESS

problem solving, communication, reasoning, pattern, probability

DOMINOES

problem solving, communication, reasoning, patterns, counting, equations, equivalence, fact strategies, mental math,

LIFE

problem solving, communication, reasoning, computation, estimation, patterns, probability

MONOPOLY

problem solving, communication, reasoning, computation, estimation, probability

SIMON

problem solving, communication, reasoning, patterns

SORRY

problem solving, communication, reasoning, computation, estimation, algebra

SPIROGRAPH

problem solving, reasoning, patterns, geometry, measurement

TANGRAMS

angles, area, classification, fractions, reasoning, patterns, ratio/proportion, size/shape/color, visualization, symmetry, computation, estimation

TIC TAC TOE

problem solving, communication, reasoning, patterns

UNO

problem solving, communication, reasoning, patterns

YAHTZEE

problem solving, communication, reasoning, number relationship, computation, estimation, patterns, statistics, algebra, probability

MANIPULATIVES THAT CAN ENHANCE MATH SKILLS - IDEAL FOR KINESTHETIC LEARNERS

CAPACITY CONTAINERS

(cups, pints, quarts, gallons, etc.)

estimation, fractions, measurement, volume

CALCULATORS

counting, decimals, estimation, fact strategies, fractions, number concepts, patterns, problem solving, whole numbers

CLOCKS

fractions, measurement, whole numbers

CUBES

area, classification, counting, equations, equivalence, fractions, number concepts, percent, probability, visualization, square/cubic numbers, surface area, geometric transformations, volume, whole numbers

DICE (NUMBER CUBES)

counting, decimals, logical reasoning, mental math, number concepts, probability, whole numbers

MONEY

classification, counting, decimals, equations, equivalence, fact strategies, money, probability, whole numbers

PATTERN BLOCKS

angles, area, classification, fractions, logical reasoning, patterns, perimeter/circumference, probability, ratio/proportion, similarity/congruence, size/shape/color, symmetry, tessellations

PLAYING CARDS

counting, number concepts, probability, (number theory)-prime, even, odd, composite, whole numbers

RULERS

area, construction, decimals, estimation, measurement, volume, whole numbers

SPINNERS

counting, fact strategies, fraction, logical reasoning, mental math, number concepts, probability, whole numbers

THERMOMETERS

integers, measurement

TWO - COLOR COUNTERS

counting, equivalence, equations, fact strategies, fractions, integers, number concepts,(number theory)- odd, even, prime, composite, place value, probability, ratio/proportion, whole numbers

77 CAREERS INVOLVING MATHEMATICS

TWO YEARS OF MATH RECOMMENDED

Agricultural Technician
Administrator Technician
Bookkeeper
Data Processor
Dental Assistant
Electronic Technician
Farm Equipment Mechanic
Industrial Administrator
Industrial Designer
Landscape Technician
Metallurgical Technician
Office Clerk
Ophthalmic Assistant
Tool and Die Maker
Welder

TWO YEARS OF MATH REQUIRED

Cartographer Technician
Commercial Diver
Computer Technologist
Dental Technician
Draftsman/Woman
Forestry Technician
General Accountant
Horticulturist
Medical Equipment Maintenance Technician
Meteorological Technician
Pharmacy Assistant
Pilot
Public Health Inspector
Public Health Nurse
Respiratory Technologist
Survey Technician
X-Ray Technician

FOUR YEARS OF MATH RECOMMENDED

Audiologist
Cartographer
Engineering Technologist
Geographer
Health Record Administrator
Lawyer
Nuclear Medical Technologist
Occupational and Physical Therapist
Property Appraiser
Psychologist
Registered Nurse
Sociologist
Speech Therapist
Survey Technologist
Teacher
Urban Planner

FOUR YEARS OF MATH REQUIRED

Actuary
Aerologist
Applied Mathematician
Architect
Astronomer
Biologist
Business Administrator
Chartered Accountant
Chemist
Computer Scientist
Dentist
Economist
Engineer
Forestor
Geologist
Graphic Artist
Interior Designer
Landscape Architect
Medical Lab Technologist
Meteorologist
Optometrist
Pharmacist
Physician
Physicist
Pure Mathematician
Statistician
Surveyor
Veterinarian

WORKFORCE 2000+ EMPLOYER MATH SKILLS EXPECTATIONS

ENTRY LEVEL

Add and subtract whole numbers
Tell time on a non-digital clock
Use standard or metric ruler
Understand paycheck elements

MID LEVEL

Add and subtract whole numbers
Tell time on a non-digital clock
Use standard or metric ruler
Understand paycheck elements
* Add, subtract, multiply and divide fractions and decimal
* Do metric conversions
* Understand weight and cube
* Plot charts and graphs

HIGH LEVEL

Add and subtract whole numbers
Tell time on a non-digital clock
Use standard or metric ruler
Understand paycheck elements
*Add, subtract, multiply, and divide fractions and decimals
* Do metric conversions
* Understand weight and cube
* Plot charts and graphs
** Add, subtract, multiply and divide percentages
** Convert fractions and decimals to percentages

** Convert percentages to fractions and decimals

*Skills required beyond entry level
** Skills required beyond mid level

Developed by the Sumter County (South Carolina Development Board's Workforce 2000+ Steering Committee

ALTERNATIVE LESSONS

TO THE

TRADITIONAL WORKSHEET

PRIME NUMBER GAME

Materials 2 cubes numbered 4 through 9
1 cube numbered 1 through 6

Multiply two cubes then add or subtract the third cube. Everyone works at the same time with or without calculators. The first person to get a prime number will write it down next to his or her name on score sheet.

Players 2 to 4

The winner will be determined by the student who has found the most prime numbers.

PROBABILITY

Level: Secondary Mathematics

Topics: Basic Counting Principle
Sample Space

Objectives: Students will find the total outcome by using manipulatives (shirt & pants cut outs). Students will find the total number of outcomes by using a basic counting principle. Students will use a tree diagram to show the sample space. Students will compute the probability of an event using the sample space.

Evaluation: The students will write tree diagrams for the remaining four shirts and list the total possible outfits. The students' final product will be checked for accuracy.

INDEX

A

B

C

D

E

F

G

H

N

O

P

R

S

T

U

V

W

X, Y and Z

AUTHOR'S MESSAGE

I hope that you have enjoyed this book as much as I have enjoyed preparing it for you. Hopefully you will now look at math with new confidence. I am available for seminars at schools, parent organizations and corporations. For further information or additional copies of *Math Attack* please write to:

Marilyn Curtain-Phillips
Math Attack
PO Box 62
Manning, SC 29102-0062

E-Mail: mathattack@yahoo.com

For more information see our web pages at:
http://www.oldmp.com/mathattack
and
http://www.amazon.com

MATH ATTACK BOOK ORDER FORM

Enclosed is my ___check or____ money order for $15.00 (including postage and handling). Please rush me *Math Attack,* I am ready to face math with new confidence.

Name ____________________________________
(Please Print)
Address __________________________________

City _____________**State** ____ **Zip Code** ______